Fantastical Learning
A Journey Through STEM
for Young Explorers
By Yanki Farouez

Disclaimer

Copyright © [2023] by Yanki Farouez

All rights reserved.

No part of this publication may be reproduced, distributed, or transmitted in any form or by any means, including photocopying, recording, or other electronic or mechanical methods, without the publisher's prior written permission.

Book Cover by Yanki Farouez

Illustrations by Yanki Farouez

[1] edition 2023

Dedication

To all the curious and imaginative young minds out there,

May this book ignite a passion for exploring the wonders of science, technology, engineering, and math. May you discover the joy of asking questions, making discoveries, and solving problems. May you never lose your sense of wonder and may the magic of STEM always guide you on your adventures.

With love and excitement for what the future holds,

Yanki Farouez

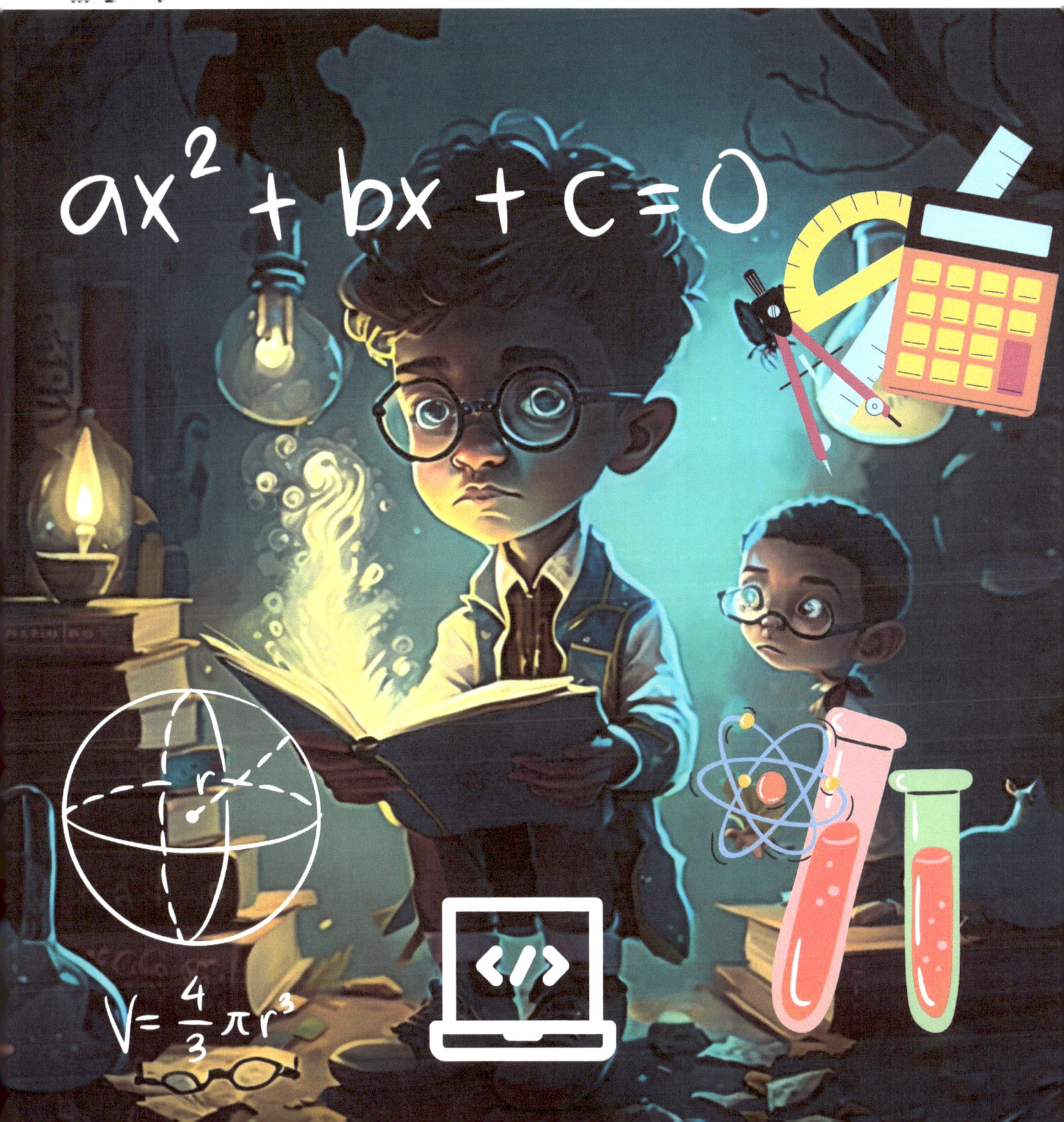

Once upon a time, there was a young boy named Max who enjoyed discovering new things. Max was always interested in how things worked and why things happened the way they did. Max came across a fantastic book one day that promised to transport him to the wonders of science, technology, engineering, and math.

Max eagerly opened the book and was transported to a fantastic laboratory full of scientific tools and experiments. There, Max met Dr. Smith, a nice scientist who exposed him to the wonders of science.

Dr. Smith demonstrated how to do easy experiments with common household ingredients such as baking soda and vinegar. Max was shocked by how the two chemicals reacted when they were mixed together, causing an explosion. Dr. Smith described the chemistry of the reaction and encouraged Max to continue researching.

Max was then transported to a technological realm. There, he met Robby, a nice robot who showed him how technology might make our lives easier and more efficient. Robby also taught Max how to write simple programs using basic code.

Max was brought to a building site after researching technology and meeting a bunch of engineers. The engineers demonstrated how they designed and constructed bridges, buildings, and other things to Max. Max was very interested in the engineering process, and he even helped build a small bridge.

Ultimately, Max was transported to a mystical math country, where he met a bunch of math wizards. Max was shown by the wizards how to solve puzzles and issues utilizing basic mathematical concepts such as addition, subtraction, and multiplication.

As he progressed through the magical book, Max found that science, technology, engineering, and math were all interconnected. He learned how to use science to comprehend the world around us, how to use technology to solve problems, how to use engineering to design and build structures, and how to use math to solve difficulties and riddles.

Max was thrilled to find out that he could use what he had learned to make his own inventions and solve problems in his everyday life. He wanted to use his new skills in coding and engineering to make a simple robot.

Max was able to make a rudimentary robot that could move and light up with the help of Dr. Smith, Robby, and the engineers. He was overjoyed with his accomplishment and couldn't wait to show it off to his friends and family.

Max's voyage through the wonderful book was over, but he knew his learning journey had only just begun. He discovered there was so much more to study and explore in the realm of STEM, and he was eager to continue his education.

Max chose to share his enthusiasm for STEM with his friends and family, and he even founded a STEM club at his school. He hoped that future young explorers would be inspired to learn and find in the same way that he had been.

Finally, Max discovered that the world of STEM was full of magic and wonder and that there were no limits to what he could discover and build. He was grateful for the amazing book that had brought him on this incredible voyage, and he knew that his passion for learning and exploration would last the rest of his life.

Thank you for being a part of this journey with me, and I hope to have the opportunity to share more adventures with you in the future.

Dear Readers,

I want to express my deepest gratitude to everyone who has picked up a copy of "Fantastical Learning: A Journey Through STEM for Young Explorers". Your support means the world to me and I hope this book has brought joy and excitement to the little ones in your life.

STEM education is such an important part of our children's future, and I truly believe that fostering a love of science, technology, engineering, and math from a young age can open up endless possibilities for their future. It was my goal with this book to introduce these concepts in a fun and engaging way, and I'm so grateful for the opportunity to do so.

I would also like to extend a special thanks to all the educators, and parents who have shared this book with their students, children, and little ones. Your dedication to fostering a love of learning is truly inspiring, and it's an honor to be a part of that process.

Thank you again for your support, and please feel free to connect with me on my author page https://www.amazon.com/author/lovereading to share your thoughts, reviews, and feedback.

Sincerely,

Yanki Farouez

www.ingramcontent.com/pod-product-compliance
Lightning Source LLC
Chambersburg PA
CBHW051954210526
45473CB00030B/2296